What does it feel like in space?
How long does it take to get used to space?
Do you look the same in space?
Does your hair float out from your head?
What kind of work did you do on Skylab?
Was it hard to work while in a space suit?
How did you drink?
How did you bathe?

All the answers to all the questions you have about living in space, including the question most often asked . . .

HOW DO YOU GO TO THE BATHROOM IN SPACE?

HOW DO YOU GO TO THE BATHROOM IN SPACE?

WILLIAM R. POGUE, ASTRONAUT

Cartoons by Sidney Harris

TOR

A TOM DOHERTY ASSOCIATES BOOK
NEW YORK

A Tor Book
Published by Tom Doherty Associates, Inc.
175 Fifth Avenue
New York, N.Y. 10010

Cartoons by Sidney Harris
Photographs courtesy of NASA

ISBN: 0-812-51728-8

Printed in the United States of America

Revised edition: August 1991

0 9 8 7 6 5 4 3 2 1

TO JEAN AND ALL OUR GRANDCHILDREN

ACKNOWLEDGMENTS

I received considerable help in the form of assistance and advice during the revision of this book. Hal Stall, Doug Ward, Jack Riley and Chuck Biggs of the Office of Public Affairs at the Johnson Space Center were especially helpful in providing photographs and technical details for answers to many questions on current NASA programs. Mike Gentry at Johnson was a great help in locating photographs and illustrations. Jim Poindexter and his space education specialists at the Johnson Space Center reminded me of many questions I had omitted from the original manuscript. I also received substantive encouragement and many suggestions from the staff that administers the NASA Aerospace Education Services Program, Dr. Ken Wiggins, Dr. Nelson Ehrlich, and Dr. Malcom Phelps at Oklahoma State University. They gave me a real morale boost by distributing the book to their aerospace education specialists that serve students and teachers in schools throughout all fifty states, the U.S. territories and parts of Canada. The index was added in response to a suggestion from teachers participating in the Alabama Space Education Workshops. I am grateful to them for a practical idea that makes the book easier to use by teachers.

My agent and good friend, Barbara Bova, exerted an enduring effort in getting the book published and contributed substantially to editing and revising the content; without her help, the manuscript would be gathering dust on my shelf. At Tor Books, Kathleen Doherty has contributed greatly to the acceptance of the book by prevailing in overcoming the trade's aesthetic objection to the title. Her persistence is greatly appreciated.

My wife Jean was instrumental in recording questions at presentations. She also provided extensive editing suggestions to clear the fog from many of my answers that were too technical for the general reader; and labored at length to prepare and revise the manuscript. She has my gratitude, appreciation and love for bearing with me during our effort to create this book.

INTRODUCTION

While serving With NASA and also since my retirement as an astronaut, I have received many invitations to speak to the public about space exploration. Whether they be civic, business /professional, church or student audiences, they are all curious and they all ask questions.

During the past twenty five years, I have spoken about space exploration to audiences in twenty-six states and eight foreign countries. Most of these presentations were made to schools and nearly all of them included an opportunity for a question and answer session. The questions asked by people in Thimpu, Bhutan were not much different from those asked by people in Witchita, Kansas. Curiosity about space flight is an international trait.

After hundreds of presentations, my wife, Jean, and I finally started keeping a log of the questions and the number continues to grow. Just when I think I've heard them all a new one pops up. However, one question dominates the list in terms of frequency an interest. How do you go to the bathroom in space? I have been asked this question by kindergarten students and I've gotten it over dessert at the Harvard Club. The wife of an eminent astronomer blasted me with the question just as I was taking my first bite of chocolate mousse. That was the last time I was surprised by this question. Believe me, I had an answer for the next several hundred times it was to be asked.

Many people have advised me to write a book about space. At first, I couldn't visualize a book that would present general and technical information in a form that would be interesting as well as informative, and rejected the idea.

Finally, it occurred to me that the questions from my audiences revealed a clear area of general interest. Why write a dry technical book or a book about my personal perspective? Rather, let the audience write the book or let their questions dictate the content. The questions themselves are interesting and reflect the curiosity of a wide age span. Some seem trivial while others require research to answer; the variety is also intriguing.

So the questions from my audiences form the basis for this book. I have tried to keep the answers brief and they are more or less the same as the verbal answers I have given when speaking.

1. How old are you?

I was born in 1930. I was forty-three years old when our mission launched to visit *Skylab,* and I celebrated my forty-fourth birthday while in space.

2. What is your academic background?

I have a bachelor of science degree in secondary education and a master of science degree in mathematics, and I taught undergraduate mathematics at the Air Force Academy.

3. Are you a graduate of a service academy?

No, I attended a civilian college and received my commission through the Air Force aviation cadet program during the Korean conflict.

4. Which branch of the military [service] were you in? How long?

I was in the Air Force for twenty-five years, nine of which were spent with NASA as an astronaut.

5. What kind of airplanes have you flown?

Over fifty types and models of American and British aircraft, mostly jet fighter aircraft, but including civilian light aircraft, sailplanes, helicopters, open cockpit biplanes of World War II vintage, and four-engine patrol bombers used by the RAF Coastal Command.

6. What kind of airplane did you fly in Korea?

The F-84G, a fighter-bomber.

7. When did you fly with the Thunderbirds? What airplane did you fly with the aerobatic team? What position?

I flew with the Thunderbirds, the Air Force aerobatic team, from 1955 to 1957. I flew as solo pilot in the F-84F and F-100C, and, for over a year also in the F-100C, I was slot pilot in the diamond formation.

8. Which do you think was riskier, flying with the Thunderbirds or going into space?

Flying with the Thunderbirds was probably riskier, but I never thought of it as being dangerous, because we practiced regularly. The most hazardous flying I ever did was instructing students in aerial gunnery training and air combat maneuvering [dogfighting]. Both of these activities involved a lot of airplanes maneuvering close together in the same airspace, and it required a lot of attention by the pilots to avoid midair collisions. Inexperienced pilots frequently maneuver in an unpredictable manner, which can create a dangerous situation.

9. Which group of astronauts were you in? When were you selected to be an astronaut? When did you leave the space program?

I was in the fifth group of astronauts selected in 1966 and left the space program in September 1977. I had been turned down on two earlier selections [1961 and 1963] but just kept sending in an application each time NASA announced a selection. I wanted to become an astronaut because it seemed to me to be the highest goal attainable for a pilot as well as being interesting and exciting work.

10. Do you still do work on space projects?

Yes. I support one of the companies working on the space station project, assisting in the design of facilities and equipment that the astronauts will use. In addition, I participate in advanced planning studies related to future missions to the moon and to Mars.

11. How old do you have to be to become an astronaut? At what age do astronauts retire?

No age requirements are stated for selection, nor are astronauts required to retire at any specified age. The average ages at time of selection are about thirty-three to thirty-four for mission specialists [technical/scientific astronauts] and thirty-five to thirty-six for pilot astronauts. The youngest astronaut ever selected was twenty-five, a geologist in the 1967 scientist-astronaut selection [sixth group].

12. Do astronauts have to be pilots?

No. Prior to the 1978 selection, all astronauts were required to become pilots and were trained in U.S. Air Force pilot training schools. Beginning with the 1978 selection, NASA began selecting in two categories: mission specialist astronaut and pilot astronaut.

13. How many crews visited Skylab? How long did the missions last? What were the names of the other men on your mission? What happened to Skylab?

Three crews [three men each] visited Skylab between May 1973 and February 1974. The first crew saved Skylab by performing critical repairs of damage caused during the launch of Skylab. They were: Captain Charles E. (Pete) Conrad, Commander; Captain Joseph P. (Joe) Kerwin, Scientist Pilot; and Captain Paul J. Weitz, Pilot. The second crew was: Captain Alan Bean, Commander; Dr. Owen P. Garriott, Scientist Pilot; and Colonel Jack Lousma, Pilot. The other men on my flight [final visit] were Colonel Gerald P. (Jerry) Carr, Commander, and Dr. Edward G. (Ed) Gibson, Scientist Pilot. All three flights set successive endurance records for space flight.

Skylab was put into a "storage" condition when we left it in February 1974, with the intention to service it [for later use] or deorbit it with a device carried by the shuttle later in the decade [late 1970s]. The choice was left open, because it wasn't known if it would be usable after being in space for so long. As it turned out, Skylab's orbit dropped faster than anticipated, and it reentered the Earth's atmosphere on July 11, 1979, scattering debris over a path extending eastward from the Indian Ocean to western Australia.

14. What do you think was Skylab's greatest contribution?

The greatest immediate contribution [1974] was the demonstration of man's capability to live and work in weightlessness for long periods [up to three months]. The medical/physiological tests and experiments were rigorously performed and are still [1991] the most comprehensive and reliable data available regarding the long-term physiological effects of space flight. The studies of the Earth and sun provided data for many years of analysis.

15. Why do you need to do astronomy or solar observations from space when it is a lot cheaper and easier to use Earth-based observatories that are already available?

When light from the stars, planets, or the sun passes through the Earth's atmosphere, it can create many problems for an astronomer. The light may be entirely blocked or screened out, or the light may be changed [distorted] to prevent accurate observations. Studies of the sun's faint outer features, the solar corona, cannot be seen at all because the daytime sky is so bright. On one orbit of *Skylab* more solar corona observations [data] were accumulated than during all previous observations by Earth-based ob-

Overhead View of Skylab Space Station Cluster with Cloud-Covered Earth in the background. This was taken from the Skylab 4 Command and Service Module (CSM) on its last orbit before returning home from its final manned mission in February, 1974. (See question 156).

servatories [worldwide]. From Earth the corona can only be seen during time of total eclipse of the sun [when the moon passes in front of the sun].

16. Aren't you really thrown back at lift-off?

There is a common misconception that the astronauts feel the strongest force effect from engine thrust at the moment of lift-off. This force isn't nearly as great as generally believed. In the older Saturn boosters [rockets], the astronauts were pressed back in the couches at just a bit more than their normal body weight; an astronaut weighing 150 pounds, for instance, would feel like he weighed 165 pounds. On the space shuttle, the same astronaut would be pressed back into the seat with a force of about 225 pounds. This is much higher lift-off acceleration than felt on the older boosters, but it isn't nearly the force felt later on during boost. On Saturn boosters the astronauts felt the most force just as the first stage of the rocket had burned up most of its fuel—about four times heavier than their normal weight. The space shuttle thrust is controlled so that the most force the astronauts feel is about three times their normal weight.

17. What does it feel like in space?

The first thing you notice when you go into space is an absence of pressure on your body. You may feel light-headed or giddy. After a half hour or so, your face may feel flushed and you might feel a throbbing in your neck. As you move about, you will notice a strong sensation of spinning or tumbling every time you turn or nod your head. This makes some people uncomfortable or nauseated. You will also have a very "full feeling" or stuffiness in your head. You may get a bad headache after a few hours, and this too may make you feel sick to your stomach.

Most all of these symptoms will go away in a few days. The head congestion or stuffiness may bother you off and on during your entire time in space. Throughout the space flight, you will feel a powerful sensation of tumbling or spinning every time you move your head too fast.

There are two things you can do on Earth to get a reasonable idea what it feels like in space. The general floating feeling is quite similar to the effect of relaxing in a swimming pool. The head stuffiness experienced in space is much like the uncomfortable feeling that one gets when hanging upside down from gymnast bars. Normally, it is uncomfortable to stay in this position beyond a minute or two because of the full feeling in the head caused by the upside-down position.

18. How long does it take to get used to space?

It takes the body about three days to adjust to the weightlessness. You will become accustomed to working in space in a few hours, but you will be learning better ways to do things throughout the mission. Even though I got sick the first evening in space, the following day, which was our first full day in orbit, I worked fourteen hours.

19. When sailors go to sea, they gradually get their "sea legs." Did you get "space legs" after being in space for a while?

We did, quite literally, develop space legs. We called it bird legs, because our legs became thinner and thinner as the weeks passed. The calves, in particular, became quite small. During the first few days in space, the legs become smaller, because the muscles of the legs force blood and other fluids toward the upper part of the body, thus decreasing the girth measurement around the thighs and calves. In addition, muscle tissue is progressively lost due to insufficient exercise. These changes produce a "bird leg" effect.

BIRDLEGS

20. What happened to your body in space?

We grew one and a half to two and a quarter inches taller. This height increase was due to spinal lengthening and straightening. The discs between the vertebrae expand and compress slightly, depending on the weight the back is supporting. Even on Earth an adult will be slightly taller [about one-half inch] in the morning than in the evening, because the discs expand during sleep and compress as you walk or sit during the day. In weightlessness, the discs expand, but they don't compress again, because there is never weight on the spine. Our space suits were custom-tailored on Earth to our height and posture; thus, they fit tighter in space because of the height increase. Also, my waist measurement decreased by almost three inches, due to an upward shift of the internal organs in the body, creating a "wasp waist" appearance.

In addition to the height increase and waist thinning, the posture changes slightly. Your relaxed body posture is semi-erect with the knees slightly bent, head tilted forward, shoulders up (similar to a shrug), and arms floating up and bent in at the elbows with hands in front at about chest height. This posture is referred to as space neutral body posture.

There is one confusing aspect of these changes. Although the body "length" is greater, the "effective body height" is actually less than on Earth because of the semi-crouch assumed by the relaxed body.

Because of the raised position of the arms, it was an effort to work for long periods at waist height as you ordinarily do; you must continually bend forward and force your arms down to waist level to do work at "table height."

Also, you can't remain in a seated position without a belt to hold your body down in the seat. The body's tendency to resume the semi-erect space neutral body posture makes it necessary to exert continuous effort with the abdominal muscles to stay "bent forward." If you flex your body to get into a chair, and then relax, you'll pop right out of the chair.

We found it much easier to pull our legs upward to lace our shoes rather than bend down. Inability to bend forward easily also made it harder to get the upper part of the one-piece space suits over our heads. We put our legs in from the opening in the back of the suit and then had to bend forward while we lifted the top half of the suit over our heads to get our heads through the neck ring. Shuttle astronauts don't have this problem, because their suits are two-piece joined at a waist ring.

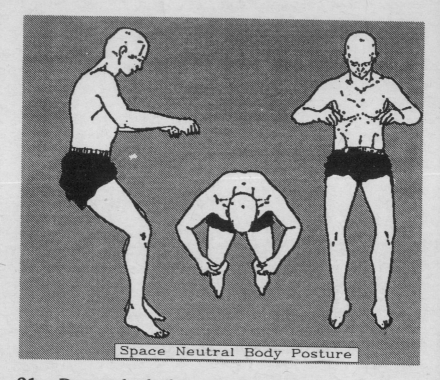

Space Neutral Body Posture

21. Do you look the same in space?

No, facial appearance changes quite a lot. I was really surprised, if not shocked, the first time I looked in the mirror; I didn't look like me anymore. Loose flesh on the face rises, or floats, on the bone structure, giving a high-cheek-boned or "Oriental" appearance. The face also looks a bit puffy, with bags under the eyes, especially during the first few days, and the veins in your forehead and neck appear swollen. After about three or four days, some of the facial puffiness [edema] and vein enlargement goes away, but your face still looks quite a bit different. Incidentally, for safety the mirrors we used were made of stainless steel instead of glass. In weightlessness, pieces of broken glass would float around and create a serious threat to the astronauts.

WE GREW TALLER

BEFORE   DURING   AFTER

22. Did you gain or lose weight?

We lost about three or four pounds during the first four days of the mission. Then we gradually regained most of the loss, so that, by the end of the mission [eighty-four days], we were just about back to our weight at launch.

Most of the early-flight weight loss is caused by elimination of body fluids. The fluid shift causes excessive fluids in the upper half of the body; the body senses these localized excesses and reduces the total body fluid through urination.

23. If you were weightless, how could you weigh yourself?

We really didn't weigh ourselves the same way we did on Earth, because a spring or balance scale wouldn't work.

However, we could determine body mass, and then convert this to an equivalent Earth weight, by sitting in a special chair that swung back and forth on springs. The time it took for each swing of the chair was measured and used by a computer to determine our body mass, or Earth weight. This device was called a body mass measurement device [BMMD].

I found the device mildly unpleasant, because the metal was cold and we used it the first thing after waking up, while we were still in our underwear. The chair was actually more like a box, and we had to get in it with our knees doubled up under our chins. This jammed-up position was required to reduce body fluid slosh and internal organ movement in our bodies during the oscillations of the chair. Minimizing this "body slosh" was necessary to get an accurate reading with the device.

24. Is it true that you lost a lot of calcium from your bones and that this may make long missions impossible?

This is partly correct. We did lose calcium from the bones, but it wasn't excessive. After *Skylab,* NASA doctors worried that this loss might become bad enough on long space flights to cause serious harm to the skeletons of astronauts. Estimates based on *Skylab* data showed that bone mass loss would be 6 to 7 percent for a year-long stay in weightlessness. Since then the Soviets have completed progressively longer missions leading up to a year-long mission aboard their *Mir* space station in 1987 and 1988. The cosmonauts exercised diligently, completing two one-hour exercise periods per day using a treadmill and stationary bicycle. It appears that this has stemmed the calcium loss.

The calcium loss is similar to that experienced by bed rest patients on Earth. Such patients show a marked decrease in the rate of calcium loss after the first few months of being bedridden. One view of the space loss is that it is also self-limiting; that is, the loss rate drops off after several months in weightlessness.

After sixteen years I have had no ill effects [skeletal problems] from a three-month stay.

25. Why don't you just take calcium supplements [tablets] to balance the calcium loss?

The bone mass loss cannot be offset [at present] by taking mineral supplements. Mineral supplements are not used by the body to replace losses in the bones and can contribute to the development of kidney stones. There appear to be three general approaches to controlling the calcium loss:

a. Exercise countermeasures as practiced on *Skylab, Salyut,* and *Mir* space stations
b. The use of mineral supplements if some technique can be developed to induce the body to use the minerals to restore bone mass [instead of dumping the supplements through the kidneys]
c. Artificial gravity countermeasures that involve periodic exposure to force generated in a rotating device called a centrifuge, or long-term exposure to the same type of force generated by rotating the entire spacecraft assembly

Despite the Soviet experience, NASA is taking a cautious approach to this problem, is studying all three options, and is planning rigorous medical and physiological tests for space station *Freedom* crew members.

26. What would happen if you got a bone broken in space? Would it knit [heal] properly?

We were trained to treat the problem by attaching a splint; we wouldn't have tried to set the bone, because we

LIVE AND WORK IN
WEIGHTLESSNESS

didn't have the training or the equipment such as X-ray machines.

The question of bone healing has not been encountered, because no crew members have broken a bone in space. Breaking a bone in space is highly unlikely. However, we need to understand more about the basic process of bone tissue restoration before this question can be answered.

Tests were performed on rats during a recent shuttle flight [January 1990], and the results indicate that an untended fracture may not heal properly. The concern is that the bone would not heal with sufficient strength to handle the loads and forces that would be encountered upon return to Earth. However, there are ways to induce proper healing even in space, such as the simulation of Earth-forces by exercises and exercise equipment.

27. Do you get tired in space?

Yes, we did get tired in weightlessness. Heavy exercise left us with a comfortable tired feeling. We also experienced a psychological weariness from rushing and from mental pressure to keep on schedule. All astronauts on *Skylab* have reported a sort of overall tiredness, a fatigue or run-down feeling, that often occurred about three or four hours after eating. The *Skylab* astronauts called it space crud. It's sort of like the down-and-out feeling you have when you're coming down with a bad cold or the flu. I still don't understand what caused this, but we learned very quickly that it was unwise to skip meals to save time. If we did, we would begin to feel bad and were much more likely to make a mistake.

I noticed one peculiar inconsistency about the space crud; I didn't develop it on space walks, even though I went for six to seven hours without eating or drinking. I don't know why the effects should be absent on space walks, unless it was because we enjoyed it so much that it helped to offset the occurrence of those types of symptoms.

28. Does your hair float out from your head?

Yes. Our hair was short enough that this wasn't a problem. It appears from mission video and photos that medium-length curled hair is also manageable, but the hair does extend out from the head more than on Earth. Long straight hair and slightly wavy hair spread out and, from the flight scenes aboard the shuttle, it looks like it could present a problem; it's easy to get things snagged in weightlessness. I noticed from studying postflight films that female astronauts with long hair use varying methods to manage their hair, sometimes letting it float free and other times rolling it into a bun or tying it in a sort of ponytail.

29. How did you breathe? Was breathing any different?

The spacecraft is pressurized or filled with air; on *Skylab* space walks the space suit was inflated by a steady flow of air supplied through a long hose called an umbilical. The continuous flow made sure there was always fresh air in the suit. Suits used by shuttle astronauts contain an oxygen supply in tanks attached to the back of the suit, and the astronauts are not attached to the shuttle by a hose or an umbilical.

Although we didn't notice any difference, tests [vital capacity] made on *Skylab* showed we couldn't breathe as deeply [inhale as much air] as during preflight tests. In weightlessness, there is a noticeable shift of the abdominal organs upward toward the rib cage. This may have been the reason it was harder to breathe as deeply.

30. How did you avoid high humidity from building up in the air, due to respiration and evaporation from the body?

The air inside *Skylab* was circulated across cold metal plates, where the water vapor condensed, and was then collected and transferred into a waste water tank. The device was called a condensing heat exchanger. The shuttle uses a similar system. On *Skylab* this system was very effective, and the air was usually quite dry, about the same as a desert climate.

However, in smaller volumes and with more crew members, humidity buildup can be a problem. During short-mission crew visits when there were six cosmonauts on the *Mir* space station, spokesmen reported an increase in the temperature and humidity. As they add more modules [volume] to the *Mir,* they expect much less of a problem; the growth modules will increase overall systems capacity, including humidity control.

31. What kind of soap did you use?

We used both bar and liquid soap [shampoo]. It was similar to Neutrogena. The bar soap had a disc of iron in it so it would stick to magnetic posts in the handwash area. Astronauts on the shuttle use Ivory soap.

32. What is [space] warp?

Space warp is the imaginary concept of science fiction writers. It is envisioned as a way of getting from one place to another without crossing the distance between them. At the present time, no one is able to achieve such a thing as space warp.

33. How did you get from one place to another inside the spacecraft?

Inside *Skylab,* we pulled ourselves along structural surfaces by using handholds [special handgrips or handles] or other parts of the spacecraft; we also shoved or pushed off from one position to float to the next location. When

Space Shuttle Extravehicular Activity (EVA) Astronaut Jerry L. Ross, during a 1986 mission, approaches an experimental tower to be used to test construction techniques in outer space.

shoving off to float across a large empty volume, we went headfirst, feetfirst, and also sideways—launching ourselves in a manner to keep from tumbling en route. We were usually able to arrive at the next location in the best position to grasp and hold fixtures. In small volumes, however, we had to be careful to avoid damage to the spacecraft when we moved around. Headfirst motion was the one used most often.

The "floors" of two of the large compartments in *Skylab* consisted of an open triangular grid or mesh; our shoes had a special triangular cleat on the soles that fitted into the triangular holes in the grid, and locked into position by a twist of the foot. These were used to hold us in position at work sites. As an experiment, I replaced the triangles with mushroom-shaped metal cones and tried to "walk" across the grid by hooking the edge of the mushroom in the grid openings. It didn't work; every time I stretched out my foot to take the next step, my foot just floated in the air and I had to exert considerable effort to get my foot down toward the floor grid. It was a lot harder than just floating across the area.

34. How did you get between locations on the outside of Skylab?

Moving around outside the spacecraft requires greater care. On an EVA [extravehicular activity, or "space walk"] transfer handholds and other aids were essential in getting from one place to another. Foot restraints were located at all planned workstations so that the astronaut didn't have to hold on to something continually in order to maintain position. Transfer handholds were often similar to the rungs of a ladder and were spaced for moving hand over hand to travel between locations.

Shuttle astronauts on EVA have other aids in addition to handholds and foot restraints. A cable is attached along one edge of the payload bay. EVA crew members can attach a tether [strap] from their suit to the wire, using a hook on the end of the tether. Thus secured, they can move the full length of the payload bay [sixty feet] without handholds. They also have a device that fits to the end of the remote manipulator system [robot arm], called a manipulator foot restraint [MFR]. When a shuttle astronaut gets into this movable foot restraint, the robot arm can move and position the astronaut for work throughout a large area around the shuttle.

For special missions, the shuttle astronauts may use a manned maneuvering unit [MMU]. The MMU is actually a miniature spacecraft that can be flown free of the shuttle by a suited crew member. [The shuttle suits include self-contained life-support systems and thus do not require a hose connection to the shuttle.] The MMU greatly increases the work capability of astronauts, enabling them to fly out from the shuttle to inspect satellites and dock [attach] to them. The control system of the MMU then can be used to maneuver the satellite back to the shuttle for repair or return to Earth for repair of analysis.

Skylab Space Walk.   Scientist-Astronaut Owen K. Garriot, science pilot on the second Skylab mission, is performing Extravehicular Activity (EVA or, "space walk") at the Apollo Telescope Mount of the Skylab space station in Earth orbit.

35. What kind of work did you do on *Skylab*?

Skylab was an experimental space station, and we operated equipment and instruments for more than fifty experiments. The three principal experiments were performed throughout the mission and occupied a lot of time each day. We had a solar observatory for making observations of the sun, cameras and other instruments for studying the Earth, and equipment for several medical experiments to determine the long-term effects of weightlessness on the human body.

We also changed filters, replaced drive motors on tape recorders, did housekeeping chores such as vacuuming filter screens, and repaired broken equipment.

36. What kinds of problems did you have in doing repair work?

Some of the repair jobs were really crude, because we didn't have the right tools or materials. Ed Gibson operated one instrument that used batteries to power a light inside to create a sighting reference on a mirror surface. The batteries kept running down and Ed had to keep installing fresh batteries. Finally, he ran out of spare batteries, so he had to dig out some other batteries, which were intended for our tape recorders, and try to make them work. They didn't fit in the instrument, so he had to run wires from the inside of the instrument and tape them on the battery terminals. It worked, but it really looked weird. The clump of batteries sort of floated on the end of the wires, and Ed finally had to tape the whole glob on the side of the instrument. Ed said we were bumping into the batteries and loosening the wires. I was surprised that the tape held the wires snugly enough against the battery terminals to make a good electrical connection. Early in *Skylab* training we had asked for a soldering iron as a part of our tool kit, but the mission planners said there was no need for one and disapproved the request. As it turned out, a soldering iron would have been very useful when Ed rigged his battery assembly.

WE DIDN'T HAVE THE RIGHT TOOLS

37. Was it hard work?

Yes. Some tasks were very difficult, but most were routine and merely took adequate time. The hardest jobs were repair and maintenance tasks that involved a lot of physical force, such as pulling or twisting.

Sometimes there was no convenient surface or handgrip to hold on to with the free hand to keep from floating or swaying when we handled equipment. A typical problem we faced was in trying to loosen or tighten screws on a large flat surface. When using tools requiring a lot of force like screwdrivers or socket wrenches, it was necessary to have your body firmly restrained or tied down before applying force; otherwise, when you applied force, your body would move instead of the wrench. For example, if you floated over to a panel, inserted a screwdriver into the slot of a screw, and twisted your wrist, the screw wouldn't turn—you would! To apply a strong twist force, you normally have to push on the screwdriver while twisting it. In doing one repair job, I found it necessary to rig pull straps for my left hand so I could pull to balance the push force of my right hand. If you're patient and aren't pressed for time, these problems become an interesting challenge. If you're trying to meet a critical time schedule, it can be very irritating and frustrating.

On the average, we figured it took about twice the time to do a job in weightlessness that it took on Earth. One particular servicing task I did was extremely hard to do even on the ground. It involved reaching into an enclosed area to disconnect and reconnect two plumbing lines by sliding metal sleeves on pipes that ran sideways to my arm direction. I only had to do it once in space, but that was enough. My legs were thrashing around as I tried to get my arms in the best position, and, when I finally finished, my wrists were scratched and cut from banging into the sheet-metal edges around the access opening. I really felt "pooped" by the end of the day. You can't "stretch yourself" like you can on Earth when you keep going to finish a long job.

38. What did you do on space walks?

On space walks we changed film magazines in the solar telescopes [seventy-five Earth pounds each], mounted instruments for solar studies, carried out a large camera to photograph a comet [Kohotek], installed and removed test samples to evaluate the effect of space on materials, photographed the outside of the *Skylab,* and retrieved experiment equipment installed by the previous crew. Repair jobs included modifying the drive mechanism of a radar antenna used for Earth studies and unjamming the filter wheel of a solar telescope.

THE SCREW WOULDN'T TURN — YOU WOULD

39. Was it hard to work while in a space suit?

Yes. It was hard work. The suit is bulky and stiff, which makes it difficult to bend or turn your body. The gloves are very thick, so you don't have much feel. Because air pressure in the gloves tends to hold the fingers out straight, it is very tiring to maintain a grip on anything. I always felt like a "bull in a china closet" when working in a space suit. After doing a lot of work in a suit, my fingertips became very sore and tender and I had cuts and burns on my shoulders from the braided metal arm support cables inside the suit. Even so, we enjoyed the space walks and looked forward to the chance to get outside. I was out on a space walk once for six and a half hours and once for seven hours.

The suits used by shuttle astronauts provide better torso mobility [twisting at the waist], the arms and legs are a bit easier to move, and the suit is easier to put on, although they still cannot don the suit without assistance. However, the gloves are not much better than the older designs; hand and finger movement still requires a lot of effort, and arm/hand fatigue is still experienced.

Currently [1991], NASA is testing two new suit designs, each designed to work at a higher pressure, with the objective to improve overall body mobility and hand dexterity [better glove]. The higher pressure is desired to reduce the time a suited astronaut must wait before entering and de-pressurizing the air lock prior to going outside into the space vacuum. In going from a high pressure to a low pressure, dissolved gases [nitrogen] in the blood tend to "fizz" out [come out of solution]; this is similar to what happens to a soda beverage when the cap or tab is removed from the container. When this action takes place in a person's body, the nitrogen bubbles can cause several problems; the one most familiar is the "bends" that scuba divers or deep-sea divers experience when they come up too fast from deep dives. To reduce this possibility, astronauts breathe 100 percent oxygen for a period of time to remove the nitrogen from their blood (a bit comes out from the lungs each time they exhale). The length of time needed to do this depends on the difference between the spacecraft cabin pressure and the operating pressure of the suit—the higher the suit pressure, the less time to get the nitrogen level down to an acceptable concentration. Cutting down this oxygen "pre-breathe" time is desired to reduce the time lost just waiting, and to enable the astronauts to get outside more quickly in the event emergency work is required on the outside of the spacecraft.

40. Would a person without legs be able to work in space?

Based on my experience, I feel that such a person would be able to work quite effectively in the weightless condition. Virtually all of the work in space is done with the arms and hands. The feet and legs are used mainly for holding you in position at a work location and for shoving yourself from one location to another. However, you also can move yourself quite easily by using arms and hands.

41. When you were on a space walk, did you work during darkness?

Yes, we normally worked during darkness. In Earth orbit at altitudes from 120 to 500 miles, the time for one complete orbit is approximately ninety minutes (slightly longer for higher orbits; slightly less for lower orbits). Generally, about fifty minutes are in sunlight and forty minutes in darkness.

We had no difficulty working in areas where lighting was provided. Some repair work had to be done in locations where no lighting had been installed because the designers had not expected a problem would occur there. In those locations we had to stop work when it got dark. Sometimes it started getting dark right in the middle of a long task and created a problem as we tried to "button up" the work area to wait for sunrise. Although darkness occurs quickly in space, we usually had ample warning, because the Earth below us got dark a few minutes before we flew out of the sunlight.

42. What did you do if your nose itched when you were in the space suit?

Not only did my nose itch occasionally, but also my ears. Because a scratch is almost an involuntary reaction,

I frequently reached up to scratch my nose and hit my helmet—which can make you feel really dumb. I scratched my nose by rubbing it on a little nose pincher device we used to clear our ears. If our ears stopped up or became uncomfortable due to pressure changes in the suit, the procedure was to press the nose against this open "V" device in order to hold our nostrils closed while we exerted a slight blowing pressure. This is a common technique used by fliers to clear their ears. If our ears itched, we just had to tolerate it. I usually tried rubbing the side of my head against the inside of the helmet, but it didn't help much. The best thing to do was to think of something else.

43. Was it quiet or noisy in space?

Sound can't travel through space, because there is no air to carry the sound waves. However, there was a moderate noise level inside *Skylab,* most of which was caused by pumps, fans, and voice chatter on the radio. We had a teleprinter which made a pecking sound similar to a typewriter—this made it difficult to sleep at times. Occasionally, small thrusters on the outside of *Skylab* would fire, which sounded like someone hammering on a large piece of metal.

The most peculiar sound we noticed was a deep rumble which occurred about every forty-five minutes. It sounded like the roll of distant thunder. Jerry asked, "Is it just me or do you hear a rumble every hour or so?" We all agreed that we heard it and finally decided it was due to alternate heating and cooling of the side of *Skylab* that faced the sun. This surface expanded as it heated up and shrank [contracted] as it cooled. The noise created was similar to the crackling sound made by a furnace or wood stove as it heats up or cools down. The total structure of *Skylab* was so large that it produced a low-pitched rumble instead of a crackling noise.

The most disturbing noise on *Skylab* was a loud squeal from the intercom system that occurred when the system wasn't adjusted properly.

Astronauts who have flown on *Skylab* and the shuttle have commented that the shuttle is much noisier. The shuttle cabin volume is smaller than *Skylab*'s, and it has more pumps and fans in a smaller area. Keeping the noise level down is a major objective of designers working on space station *Freedom*.

44. What kind of tools did you have?

Because of the long missions scheduled on *Skylab*, it was thought that we would need an assortment of tools to repair breakdowns in equipment. We had a rather complete set of light tools, and we used most of them at one time or the other. Most of the tools were bought at hardware stores, but some were specially made for anticipated repair work. The tools included various types of screwdrivers and pliers, socket wrenches, and torque wrenches. Special tool kits were also provided for repair jobs on space walks. Each astronaut had pockets in his trousers to carry a Swiss army knife and a pair of surgical scissors, which were used frequently for minor repair work.

Many new tools have been added during the shuttle area. Portable power tools have greatly reduced the time and effort required to remove and replace fasteners (screws, bolts, nuts). Socket tools now have special locks to prevent loss of sockets during EVA work [space walks]. The manipulator foot restraint [MFR] attached to the RMS [robot arm] can be fitted with a portable workstation. The portable workstation is an open framework in front of the EVA crew member when his feet are in the foot restraints of the MFR. It serves as a handhold and also a tool "caddie" to store the tools and keep them in easy reach.

45. Did you ever lose anything?

Yes. Several items were lost and never found. Frequently, our tableware, usually a knife, would get knocked off the magnetized surface on our food trays. The airflow in *Skylab* would usually carry the items to a filter screen in the air duct system, where they would stick due to the slight vacuum. This was the first place we looked when something was missing.

One day, when I whirled around to get a camera to take a picture of Hawaii, my eyeglasses flew off. I heard them bouncing around through the experiment compartment as I was taking the picture, but when I went to get them, I couldn't find them. Three days later, Dr. Gibson found them floating near the ceiling in his sleep compartment.

I had a spare set of half-specs, granny glasses, which I used until Ed found my bifocals. I didn't like the half-specs because the straight earpieces allowed the lenses to float up off my nose and bob up and down in front of my eyes. It was very distracting when I was using both hands to do a job.

Current astronauts have elastic eyeglass restraints to prevent this "bobbing" problem. Also, astronauts now have the option of using contact lenses. [See Question 68.]

46. How did you clean the spacecraft, or did you have to clean the spacecraft?

As on Earth, a lot of trash accumulated during the day [food packaging, tissues, wet wipes, dirty towels and washcloths]. Most of this was immediately shoved through a push-through slot into a waste container. However, bits of skin, fingernails, hair, food crumbs, odd pieces of paper, and the like tended to drift around and eventually were sucked up against air filters. We used vacuum cleaners to clean off the filters, and that took care of most of the problem. The worst mess was in the area where we ate. Small drops of liquid from our drinks and crumbs from our food would float around until they stuck on the wall or in an open grid ceiling above our food table, and it became quite dirty. Although we could see into this ceiling area, we couldn't get our hands in to wipe it clean, so it became progressively worse throughout the mission. Near the end of the flight, it began to look like the bottom of a birdcage. I just stopped looking at the ceiling after a while, because it was such a mess.

Every two weeks we had to wipe down the walls and surfaces of the toilet with a biocide [disinfectant] to prevent a buildup of microorganisms [germs, mold, etc.]. Periodic cleaning of this type will be required for space station *Freedom,* to prevent a gradual buildup of biologically active contamination. This will be a time-consuming procedure but will be essential to preserve a healthy environment for the space station astronauts.

Shuttle astronauts aren't required to do this, because the missions are short and the ground crews take care of it between missions.

47. Did you write? How did you write?

I found that a lot of writing became an irritating task in weightlessness, particularly when it had to be done on narrow strips of paper that came out of our teleprinter. This paper was just a bit wider than the paper used in cash registers. It tended to curl up and was hard to hold steady on a flat surface. We often used the food tray tops, our dinner table, as a desk, and we had to exert effort to bend forward to get into a good position to write. Also, we had to hold downward force to keep our hands and arms down on the table while writing. Other astronauts didn't seem to mind this as much as I did.

The teleprinter on the shuttle uses roll paper eight inches wide. Normal clipboards can be used to hold the paper for easy reference and provides a hand rest for making notes.

On space station *Freedom* much of the information will be displayed electronically on flat panel displays of portable workstations that can be connected to data ports located throughout the various modules. However, there will very likely still be a need to use "hard copy" [paper] for various purposes.

48. Did you use pen or pencil?

We used mechanical pencils, pressurized ballpoint pens that assured ink flow in weightlessness, and felt-tip pens. The ballpoints and mechanical pencils worked well, but the felt-tips dried out very quickly in the low humidity of *Skylab*, so they weren't of much use.

Pencil leads did break off, but they didn't cause any problems, even though it would have been possible to get a piece in the eye—or inhale it.

49. How did you keep a book open to the right page?

This occasionally became a real problem. Most of our flight documents were printed on stiff paper and held together by metal rings which could be opened to insert or remove pages. We had clips to hold the books open to the right page, but they didn't attach very tightly and occasionally they would pop off and the whole book would fan open—costing us considerable time in relocating the right page. The covers were made of extra heavy paper to make it easy to find the front of the book and the index.

When reading a book from our personal library, which was mostly paperbacks, we held the book with one hand, with our thumb in the open crease, and dog-eared the pages to mark our place.

Shuttle astronauts have CRTs (functionally the same as TV picture tubes) which display text, graphic and video pictures. They also use conventional hard-copy [printed] checklists similar to the books described above.

On future spacecraft like the space station, most of the text, graphics, and imagery data will be presented electronically on flat panel screens that are associated with fixed and portable workstations. The portable units can be connected to data ports located throughout the space station modules and will provide the crew members with the data they need at most all work locations. However, they will probably still have hard-copy checklists for safety-critical tasks as backup in the event of data system failures.

WATER IN A DRINKING GLASS WOULD TEND TO CRAWL UP

50. What would happen to water on *Skylab?* How did free water behave in weightlessness?

In weightlessness, water and other thin liquids must be fully enclosed in a container to prevent them from spilling and floating around. If you could get water in a normal drinking glass, it would tend to crawl up the inside surface, over the edge, and down the outside of the glass. Free water droplets become spherical or ball-shaped. Large drops or balls of water quiver and jiggle like gelatin as they float about. On *Skylab* we performed many science demonstrations with water drops.

One of the most fascinating effects with drops of water was created by injecting air in large water drops, using a hypodermic syringe. Starting with a drop of water about two inches across, I injected air into the center—it became a hollow ball. I tried to make it larger by squirting more air into the center but missed the center and injected it into the water shell surrounding the hollow core. It formed a second hollow ball joined to the first with a flat surface between them. After a third injection it looked like a Mickey Mouse head. They were really water drop models of complex geometric shapes. Later on I regretted not taking pictures of them.

51. Did you bring food, clothes, etc., with you when you went to visit *Skylab?* What kind of food did you eat?

Skylab contained all the food, water, clothes, and repair parts for the three planned missions. Originally, our mission had been planned for fifty-six days, so we brought a twenty-eight-day supply of food (food bars, some freeze-dried foods, and drinks) to supplement the fifty-six-day supply already on board *Skylab.* We also brought enough underwear and socks for the extra month.

In addition to these supplies for the scheduled eighty-four days, we had a two-week supply of food and water stashed in our command module. This supply would have allowed us to wait two weeks for rescue if we had had to make an emergency undocking from *Skylab,* and then had a problem with the command module. The command module, our "ferry spacecraft," was a modified *Apollo* spacecraft, and it had been originally designed for a two-week stay in space. After being docked to *Skylab* and virtually inactive for twelve weeks, we weren't sure what kind of problems might be encountered after we undocked.

Each astronaut selected his menu items from a shopping list of seventy-two food items prepared by commercial companies and NASA dieticians. We had dehydrated vegetables, scrambled eggs, and spaghetti. After adding water and heating them, most were quite good. We had canned puddings, fruit [peaches, pears], and dried fruit [apricots]. In addition to freeze-dried food such as salmon, for the first time in space we had frozen food, which included steak, prime rib, pork, and ice cream. We had no bread or milk. We also had a wide variety of drinks, which included orange, grapefruit, strawberry, cherry, grape, and coffee and tea.

The only food that was a disappointment was the chili. I was really looking forward to having it, but the oil separated from the meat and sauce, and it looked very unappetizing when we opened the can. I stirred it up as much as I could and jammed my crackers into the can before eating it.

Shuttle astronauts have over one hundred food and drink items from which they can select their food. An average daily diet is about three thousand calories, consisting of 20 percent protein, 50 percent carbohydrates [starches], and 30 percent fat.

Space station crews will include astronauts from many countries of Western Europe, Japan, and Canada as well as diverse regions of the United States. The food selections will have to accommodate the tastes and preferences of astronauts from many different cultures. For instance, the typical Japanese diet is higher in protein and lower in fat than a typical American or European diet.

52. How did you keep frozen food frozen?

There was a food freezer on *Skylab,* which was kept below freezing by coolant chilled in a radiator on the outside at the rear of the space station. We had the same problem with frost inside the freezer as you have here on Earth, and we had to remove the frost frequently. We used wet cloths to melt the ice around the door and on the inside. It was a slow, unpleasant job, and we usually took turns during the process, because our hands got so cold.

The shuttle does not have a food freezer, but the space station will have a refrigerator. They will work like home appliances and have an automatic defrost feature.

53. How did you cook your food?

We didn't have to cook our food, because it was already cooked, but we did warm solid foods like precooked meats and vegetables in their metal containers by placing them in food tray cavities that were warmed by electrical heating elements.

Skylab had a hot water system, and coffee or tea could be prepared easily and quickly.

The shuttle astronauts use convective ovens, which take about a half hour to warm their food.

The space station will have two microwave ovens to heat their food, so the warming time will be much less, a matter of minutes.

54. How do you keep food on your plate?

We didn't have plates. Our food came in cans and plastic bags that fitted into cavities in our food trays with enough friction to keep them from floating out. We used a fork and spoon to get the food from the containers and a knife and fork to cut the solid meats.

There was a thin plastic cover over most of the canned food. We cut a crisscross slit in the plastic and fished the food out with a spoon or fork. The natural "stickiness" [surface tension] of the food, and the plastic cover, held it in quite well. Occasionally, little bits of food or meat juice would float out as we took a bite. We would dab at it with a tissue as it floated above the table, and we got most of it. Unfortunately, we didn't get it all, and the droplets would usually float up, due to airflow, and stick to the ceiling. Thick soups and ice cream tended to stick to the spoon, so you could eat them normally, as long as you didn't make any abrupt movements.

The system used by shuttle astronauts is similar: their food trays are portable and the crew usually attach them to the tops of their thighs while they eat. The space station will have tray tables similar to *Skylab:* the trays will be plastic, similar to the shuttle trays, and will have metal embedded in the tableware slots so that magnets in the tableware handles will keep the utensils in place.

55. Did you have catsup and mustard?

Yes. It was in little plastic sacks similar to the kind you use at a fast-food restaurant. We also had hot sauce, liquid pepper in restaurant-type squeeze bottles, and horseradish, which we mixed into a paste and spread on our meat. The liquid pepper was especially good and had a full, fresh flavor.

We had salt water in a dispenser that looked like a hypodermic syringe with a plastic nozzle. We squirted the salt solution directly on our food.

The first *Skylab* crew had no condiments at all. The Commander, Pete Conrad, really blasted the planners when he got back and raised such a fuss about the bland, yucky-tasting food that condiments were finally added. The second crew tried regular ground pepper and salt, but they didn't work too well. The salt and pepper floated around and caused a lot of sneezing. By the time we launched, the dietician had worked out a good scheme for dispensing a wide selection of condiments. Pete Conrad really did us a great service by insisting on the addition of condiments.

Shuttle astronauts have a wide variety of single-serving pouches and may select from salt [water solution], liquid pepper, catsup, mustard, taco sauce, Tabasco sauce, and mayonnaise.

56. Did you have recycled water?

No. *Skylab* carried a total of about one thousand gallons of water for drinking and bathing. If *Skylab* had been designed for repeated visits over several years, then recycling of water would have been practical. The simplest recycling system is to recover the water from the spacecraft atmosphere. This water would come from exhaled air and moisture evaporated from the skin—i.e., sweat. We actually removed this water, but we didn't use it for anything. It was collected in a waste water tank. The Soviet cosmonauts started recycling and using this water in their *Salyut* space station in the early 1970s. A more complete water recycling scheme would also include reprocessing [recovery] of water from all body waste [urine and feces].

No water recycling is required for the shuttle because of the short flights. The system would weigh more than the water recovered and would be uneconomical.

The water recovery [recycling] system designed for space station *Freedom* will involve reclaiming water from as many sources as possible (water vapor in the air, urine and feces). Rigorous tests will be made to verify that acceptable standards of quality and purity are achieved for each recycling source.

If this sounds offensive, just remember that the Earth is a closed life [ecological] system. Something has to happen to all the water in animal life waste. It is recycled through our natural system through evaporation and subsequent rainfall, if you're lucky. Some of it isn't even recycled through the natural system if your city water comes from a river downstream of another city using the river as a sewage dump.

On moon bases or on a mission to Mars, water will be a precious resource. Recycling systems will be a necessity.

57. How did you drink? What did you drink? Did you have alcoholic beverages on board?

A water dispenser similar to a water gun was used to take a drink by holding the nozzle, or point of the dispenser, in the mouth and squirting the water directly in. We had a wide range of beverages: coffee, tea, orange, grapefruit, strawberry, and grape. They were prepared by forcing water from hot or cold water dispensers into a plastic container containing a mixture which was dissolved by the water—we usually shook it up to mix it. The plastic container squeezed up like an accordion and had a valve on the nozzle to keep the liquid from leaking out. To drink from these containers, we put the nozzle in our mouths, opened the valve with our teeth, and squeezed the bag to squirt the drinks into our mouths. The system worked quite well except for the air we swallowed while drinking.

Shuttle astronauts use a system similar to *Skylab*'s. Their drink containers are better designed and they use straws to drink, compressing down the tops of the beverage containers as they sip the drink. They still have the problem of stomach gas [described in Question 58]; some think it may be due to gases dissolved in the water used to make the drinks. As of 1990, milk will be available for use on the shuttle. The bacteria in the milk is killed using a high-temperature sterilization technique developed by Utah State University. This gives the milk a long shelf life without refrigeration.

Currently, no alcoholic beverages are permitted on U.S. space missions.

58. Did you have trouble swallowing?

No. We had no trouble swallowing, but there was one bad aspect of swallowing drinks from the plastic drink containers. I think it bothered me more than the others. When I drank from the plastic squeeze-drink bags, I tended to swallow a lot of air with the liquid. This caused an uncomfortable pressure in my stomach, which normally would be relieved by burping or belching. But—in weightlessness—the contents of the stomach don't settle; they coat the stomach more or less uniformly. So, if you burp, you stand a very good chance of regurgitating. The gas pressure in the stomach is unpleasant, but the consequences of burping are even worse. I think I only burped twice in eighty-four days. Once my exercise period had been scheduled right after breakfast, and I had only been pedaling the bicycle a short time when I got this strong desire to burp. I fought it, but it happened anyway. I gritted my teeth, swallowed it, and kept right on pedaling.

59. How did you wash dishes?

Because we ate directly from plastic bags or cans, the only things that required cleaning were our tableware and food trays. These were wiped with tissues soaked with a mild disinfectant. The cans were crunched flat with a special food-can crusher and placed in a bag for disposal. We didn't throw anything into space.

Shuttle crew members use a similar system. A dishwasher is planned for the future space station.

60. What did you do with the trash?

Skylab had a large tank [two thousand cubic feet] which was used as a trash-disposal volume. We compacted our garbage as much as possible, placed it in a special bag, and forced it through a large tube into the special tank below the floor. The tube contained an air lock chamber to prevent loss of air when we opened the hatch to the tube. The assembly was called a trash air lock [TAL].

The lid on the trash air lock began to cause difficulties on the second *Skylab* mission. The hatch became more and more difficult to latch in the closed position. On our mission, the problem became worse, and we were very concerned, because it was essential to get rid of the biodegradable garbage and waste [food residue and urine bags]. We finally worked out a system whereby Jerry Carr would load the trash bag in the bin of the trash air lock and I would float above, holding on to the ceiling. As he closed the hatch, I would pull myself down sharply and stomp on the hatch lid while Jerry closed the locking lever. It sounds like a barnyard procedure, but it worked.

All trash and waste collected on the shuttle is returned to Earth. On space station *Freedom* the plan is to use a trash compactor to reduce the trash volume and store it for later return to the Earth. It will be returned in a large freight carrier called the logistics module, which will be hauled up and down by the shuttle.

WHAT DID YOU DO WITH THE TRASH?

61. Why didn't you just dump it out into space? Wouldn't it just burn up during reentry?

Dumping trash out into space is irresponsible, because it adds to the space debris [junk] already contaminating the region used by manned spacecraft. Ejected trash will remain in orbit for a long time, and there is a reasonable chance that it [dumped waste] will collide with the spacecraft on a later orbit.

62. How did you wash your dirty clothes?

We didn't. When they got dirty, we threw them away with the other trash. All trash was put into the large waste compartment at the rear of the space station. [See Question 60.]

Shuttle astronauts return their clothing, which is cleaned for reuse.

The space station will have a washer for clothing and bath linen. Although it is difficult to design a washing machine for use in space, studies show that it will save weight by reducing the quantity of clothing, towels, and washcloths required during resupply from Earth.

63. What kind of clothes/underwear did you wear?

We wore trousers, T-shirts, and jackets. The trousers had several pouch pockets with zipper closures or flap covers held closed by Velcro. The zippers and flaps were useful to prevent things from floating out. However, if you had a lot of things in a pocket, you never knew what would come out first when you reached inside. Our underwear was commercially manufactured briefs and T-shirts.

Shuttle astronauts have a wide range in their clothing selections: one-piece coveralls, shorts with T-shirts, jackets, etc. Most seem to prefer working in stocking feet on the shuttle. Before flight, the shuttle astronauts select their flight clothing at the Flight Equipment Processing Facility located near the Johnson Space Center at Houston, Texas. This facility assembles the food, clothing, space suits, cameras, special instruments, calculators, tools, etc., that are needed for each flight.

WHEN CLOTHES GOT DIRTY, WE THREW THEM AWAY

64. Did you have space pajamas? How did you get dressed, put on your socks and shoes?

We slept in our underwear. After we "weighed" ourselves each morning, we slipped on the *Skylab* T-shirts and trousers, and then floated into our shoes, which were left attached to the floor during the night. Because it was difficult to bend forward, we pulled our legs up to put on our socks and tie our shoes.

65. What would happen if your glove came off your suit during a space walk?

This is very unlikely to happen. The gloves were attached with a double locking mechanism, and it was easy to check that they were on properly (similarly with the helmet). It required a concentrated effort to unlock and remove the gloves and helmet.

However, if a glove came off, all the air would leak out and the astronaut would die.

66. How did you keep from getting too hot or too cold on a space walk?

We wore water-cooled one-piece long johns, called an LCG [liquid-cooled garment]. Cool water was circulated through plastic tubing throughout this mesh garment to remove body heat. We could control the water flow rate to regulate the temperature.

The shuttle astronauts use an improved version called an LCVG [liquid cooling and ventilation garment]. In addition to the liquid cooling feature, the LCVG includes air ducts to provide ventilation to the arms and legs. Previously, the suits included ducts for circulating air to the limbs; the LCVG has enabled simplification of the suit design.

67. Did you wear glasses?

Yes, I took two pairs with me. I'm farsighted and needed glasses to read small print and instrument displays. I didn't wear glasses in the space suit, because we didn't have a good eyeglass retention system; wearing glasses for close work ruined my distant vision, which I needed when transferring equipment to the other crew member.

68. Are astronauts allowed to wear contact lenses?

Yes, they can safely use contact lenses in space. Crew members desiring to use contact lenses during space missions are encouraged to start using them during training at least one month prior to launch, and are required to carry eyeglasses as a backup.

NASA flight physicians recommend soft contact lenses for astronauts not requiring a correction for astigmatism, and the gas-permeable types for those that do need an astigmatism correction.

69. Did you just float around when you slept?

No. One member of our crew tried this once, but it didn't work too well because he drifted around with the airflow and kept bumping into things. We slept in sleeping bags supported by a tubular metal frame that was strapped to the wall of our individual sleep compartments. We slipped into the sleeping bag feetfirst through the neck holes. There were arm slits in the bag, so we could reach out. It had straps on the front and back that we could tighten to hold us in a steady, snug position, and there were extra sleeping bag wraps that could be zipped on for greater warmth. Airflow, light, and temperature could be controlled in each sleep compartment.

Astronaut William R. Pogue in a Spacesuit. Note the bulkiness of the suit and the thickness of the gloves. The large unit on the front of the suit is a pressure control unit (PCU) that regulates pressure in the suit and the flow of water through the Liquid-Cooled Garment underneath. (See question 66).

Shuttle astronauts also use sleeping bags, which they position throughout the shuttle compartments. A few have found it more convenient to sleep while strapped in a seat on the flight deck [commander and pilot seats]. Some have said they couldn't get to sleep in the middeck [lower deck] of the shuttle because of the noise from cooling fans.

When assembly is completed, the space station will have individual crew quarters about 50 percent larger than *Skylab*'s [135 cubic feet in the space station]. Although this may seem small (slightly larger than a telephone booth), it will be arranged to permit dressing/undressing in privacy and will enable the astronauts to view video, operate computers, etc., within their private quarters.

70. How long did you sleep?

About six hours was all we needed, because we weren't using a lot of physical energy performing our tasks in weightlessness. We got wake-up calls at 6:00 A.M. central standard time.

71. Was the sleep restful, the same as on Earth?

Yes, but I think there is a difference. Tests made on *Skylab* showed that there is a change in the time you spend at the different levels of sleep. Many times the sleep was fitful because of work difficulties. Also, some astronauts have been bothered by a peculiar effect known as head nod. During full relaxation in sleep, the head devel-

Sleeping on the Skylab   Scientist-Astronaut Owen K. Garriot, science pilot on second Skylab mission, sleeps in the standard Skylab sleeping bag. (See question 69).

ops a nodding motion. This nodding motion is thought to occur as a result of blood pulsing through the large arteries in the neck. Occasionally, astronauts have been awakened by nausea symptoms, which they blamed on the head nod. Others have noticed the head nod but did not feel any ill effects from it.

72. Did you snore? Did anyone snore on *Skylab?* Did snoring bother you or keep you awake?

No. As far as I know, no one snored on *Skylab*. In weightlessness the position of the soft palate in the upper throat doesn't change with the body position, which is probably the reason people don't snore in space. Most snoring occurs when a person lies on his/her back, thus causing the soft palate to hang down and vibrate during breathing.

73. Did you have to do exercises?

Yes. Since it doesn't take much physical effort to move around in space, an astronaut must exercise regularly to prevent the muscles from getting weak. We were given about one and a half hours a day to exercise.

I normally spent half an hour on the stationary bicycle, fifteen minutes using spring and pulley [reel-type] exercisers, and ten minutes walking on the treadmill. On our day off, we sometimes skipped all exercises except the treadmill. We usually listened to music, using a stereo headset, while we were exercising on the bicycle, to help pass the time.

One time I was playing a new tape from Joe Kerwin's selection and really pumping hard as the work load increased at the end of the workout. I was tiring fast and wondering if I would be able to finish when the overture from *William Tell* started. It really gave me a shot of energy and I finished with power to spare. I was really surprised how much the music affected my performance.

As the Soviet missions increased in length [*Salyut* and *Mir* space stations], they refined their exercise programs based on their postflight evaluations of the physical condition of the cosmonauts. They maintain that it is essential to exercise twice daily for a one-hour period. The two cosmonauts that completed the 366-day mission in December 1988 were reported to be as fit as cosmonauts completing earlier missions of shorter duration.

74. What happens to sweat?

We got a lot of sweat on our backs when we pedaled the bicycle. It didn't drop off like it does here on Earth. The sweat on the back collected in a large puddle. By the end of half an hour of exercise, the puddle was as large as a dinner plate and about a quarter of an inch deep. It sort of slithered around on our backs as we pedaled the bicycle. When we were done, we had to move very carefully to avoid slinging off a large glob of sweat. It could have stuck to the walls of the spacecraft or onto equipment and caused problems. We used an old towel to mop the sweat off our backs before bathing.

Exercise really builds up a laundry pile; we started out exercising in T-shirts, but we didn't have enough for each day, and one use was enough to ruin the shirt. Eventually, we exercised wearing only our shorts. Also, the air circulation wasn't adequate around the stationary bicycle, so we used a portable fan to cool us during the heavy exercise.

75. How did you keep from floating around while exercising?

Our shoes locked into the pedals of the bicycle, but this didn't take care of the entire problem. We needed something to hold our bodies down, because we tended to float off the seat. We finally held our heads against a makeshift pad mounted against the ceiling to balance the up-force caused by pushing down on the pedals. For the spring and pulley exercisers, we locked our shoes in the floor to hold us in position while we exercised. The treadmill had a harness that held us down against the walking surface. The harness included strong elastic sections that pulled us down toward the surface with a force roughly equal to our Earth body weight.

The design of exercise restraints has been greatly improved since *Skylab*. The exercise programs on the shuttle missions have been intermittent because of the short, work-filled flight schedules. From video of their flights, the restraint system for the treadmill appears to be quite good.

Space station *Freedom* will have a variety of exercise devices: ergometer [stationary bicycle], treadmill, rowing machines, and exercisers simulating cross-country skiing.

76. Isn't it [exercise] boring?

Yes. Video displays similar to those available with commercial exercise units have been suggested for the space station. I think they would be very effective to reduce the boredom during exercise. I also think music helps, particularly music with a tempo matched to the exercise.

77. How did you go to the bathroom?

On *Skylab,* for the first time in space, we had a separate room for a toilet called the waste management compartment. A funnel-shaped device was used to collect the urine. Air was drawn through the funnel to make sure the urine was pulled into the collection bag inside the device, and this bag was changed daily. A commode, or potty, was used for solid waste collection. It was mounted on the wall (remember, there is no up or down in space) and was lined with a porous bag that was replaced after each use. Air was drawn through the bag to settle the waste.

The bag containing the solid waste was removed after each use and dried in a heat/vacuum chamber. All solid waste was dried, stored, and returned to Earth for medical analysis. Also, each day a small sample of urine was taken and frozen. It too was stored and brought back for analysis.

The toilet seat was made of a plastic-coated, stiff cushion material. A seat belt had to be used to keep the user's bottom from floating off the seat. Proper use of the toilet was essential if one wanted to avoid losing friends.

Because of their recessed plumbing, women have a special problem urinating hygienically in weightlessness. To solve this problem, NASA studied the issue in detail. This involved the photography of the urination function performed by a group of women volunteers. Based on their data, NASA developed a unisex toilet which is used on the shuttle. The unisex toilet consists of a potty seat similar to the *Skylab* commode, together with a urine collection device located near the front of the toilet seat.

In older spacecraft not furnished with a special toilet, the provisions were much cruder. Liquid waste was collected by the same method used on space walks. [See Question 79.] The urine was stored in a collection tank.

Solid waste collection was much more difficult. We used a fecal collection bag, about eight inches across, with an adhesive ring surface around the top. The user attached this bag to his bottom. After use, the solid waste was treated with chemical tablets to kill bacteria, and the bag was placed in a waste container.

This technique wasn't foolproof. Occasionally, fecal matter inadvertently floated free, unobserved by the user, and later drifted into view. Usually, no one would admit responsibility, and the event was cause for a lot of ribald comments. I can personally attest that fecal and urine spills can break the monotony on even the dullest days in space.

78. Have there been failures of the shuttle toilet?

Yes, there have been failures in the solid waste and the urine waste collection systems. When the solid waste collection system fails, the astronauts resort to using *Apollo*-type fecal collection bags. [See Question 77.] Each shuttle mission carries two bags per day per crew member, with enough for a two-day mission extension. For backup urine collection, the crew can use commercial adult diapers, which are very effective in absorbing the urine and protecting the skin from continued contact with moisture.

NASA is developing an improved toilet for the shuttle, and if it works properly, it will probably be used by the space station.

Space Shuttle Toilet   This Waste Management Compartment was the first unisex bathroom in space, designed to meet the needs of both male and female astronauts. (See question 77).

79. How did you go to the bathroom on a space walk?

We had two devices to wear under our space suits: a UCD [urine collection device] for urine and an FCS [fecal containment system] for solid waste. The UCD was attached to an undergarment with Velcro and connected to the body by a rubber sleeve containing a check valve to prevent urine from leaking back out of the bag.

The FCS was a tight-fitting, thigh-length trunk made of thick fabric. If required, the astronaut would defecate directly into the FCS. I know of no one who ever used this system, but it was good that we had it available, because we spent several hours out on space walks during *Skylab*. When we returned from space walks, we removed the waste collection devices and transferred the contents to the appropriate containers.

At the present time [1991], shuttle astronauts have several options for urine collection on space walks. Male crew members may use the UCD or commercial adult diapers. Female crew members may use the commercial adult diapers or a garment called a DACT [disposable absorbent containment trunk]. The DACT is similar to the adult diaper; it is lined with a one-way transmission layer that conducts the urine to a superabsorbent material that is capable of holding over a quart of liquid.

For launch and entry, the same options of urine collection are employed.

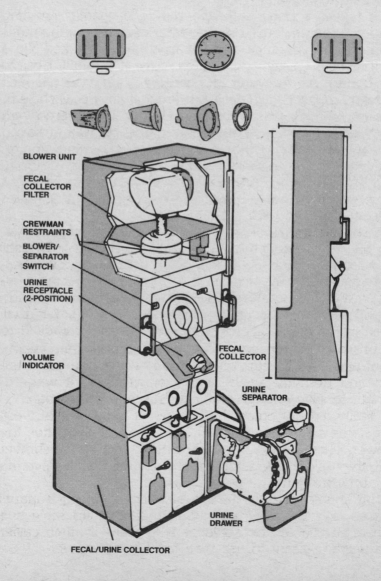

BLOWER UNIT

FECAL COLLECTOR FILTER

CREWMAN RESTRAINTS

BLOWER/ SEPARATOR SWITCH

URINE RECEPTACLE (2-POSITION)

VOLUME INDICATOR

FECAL COLLECTOR

URINE SEPARATOR

URINE DRAWER

FECAL/URINE COLLECTOR

80. How did you bathe?

We had to bathe just about every day, because we got very sweaty during exercise. On workdays, we took a sponge bath, using a washcloth, soap, and water; on our days off, once a week, we had about half a gallon of warm water for a shower.

To take a sponge bath, we started by gently squirting water on a washcloth from the water dispenser in the bathroom. The water stuck to the washcloth and looked like a thick layer of gelatin; we had to move it carefully over to our bodies. As the water touched the body, it would stick and spread over an area a bit larger than the washcloth. The entire body was wetted this way, then lathered with soap. Then as much soap lather as possible was removed with the washcloth, which could be wrung out in a special cloth squeezer. Next, water was again spread on the body and again mopped up, until the soap was removed. A towel was then used to dry. It took about thirty minutes to take a sponge bath.

A shower also took a long time—about half an hour. We had a zero-gravity shower stall, which was a circular sleeve, about three feet in diameter, with a stationary bottom attached to the floor and a circular top mounted on the ceiling. The sleeve's wall surface was fastened to the top when ready to shower and fully enclosed the user. Once inside the shower stall, a spray nozzle was used to squirt water on the body, and a vacuum cleaner attachment was used to suck off the soapy water both from the skin and from the walls of the shower stall. It was important to save enough water for rinsing off the soap.

I really did not enjoy the shower. It took a lot of work to get the equipment set up, and I got chilled after the shower. The air was so dry that when I opened the shower stall, the rapid evaporation caused uncontrollable shivering for about a minute.

The shuttle doesn't have a shower, so the astronauts take sponge baths. Space station *Freedom* will have two showers available; the objective is to make it much easier and more pleasant to use than the *Skylab* shower.

81. When you didn't bathe, did you stink?

Yes, and it's not very pleasant. One day I was running short on time and skipped washing my hair after exercise. About two hours later I was operating the solar telescopes and began noticing an unpleasant odor. I thought it was food residue. Ed Gibson frequently snacked during his tour at the telescopes, so I looked around for old food packages but couldn't find anything that might have caused the odor. The odor persisted and soon it became obvious that it was my own body odor. It was clinging around my head like a cocoon of smelly air. There is no convection in weightlessness [warm air around the body doesn't rise]. There wasn't good air circulation in this location, so I was being enveloped by my own body odor.

82. How did you shave?

We had commercial twin-blade razors, brushless shave cream in a tube, and also a wind-up rotary mechanical razor. I tried the wind-up razor but found it to be very poor. It pulled. I shaved with the blade razors for about two weeks, then stopped shaving and grew a beard. The blade razors were only good for one smooth shave, probably because there was no good way to rinse the shaving cream and whiskers from under the blades. We wiped the razor off on a washcloth and then rinsed the washcloth by squirting water on it and wringing the cloth in a washcloth wringer. It took about fifteen minutes to shave; so, when I stopped shaving, it freed up some valuable early morning time.

Shuttle astronauts have a choice of three types of commercial shaving cream.

83. Did you age less on your space journey? How much?

According to the theory of relativity, time flows at different rates under different conditions. Time passes more slowly: (1) for a person in a higher gravitational field than he normally experiences, or (2) if the person is under accelerated motion, or (3) if the person is traveling at a very high speed—near the speed of light. An astronaut experiences some of these effects, although to a very slight degree. In orbit, an astronaut is under a *weaker* gravitational pull than at the surface of the Earth; this speeds up time and therefore aging. But an astronaut experiences accelerated motion during launch and reentry, which slows down time and aging. In orbit an astronaut is traveling at about

WE HAD DIFFICULTY HEARING
EACH OTHER

five miles per second, which is far too low to cause a relativistic decrease in the aging rate. I don't know what the final result is, because the changes are too small to be determined by the spacecraft instruments.

84. Can you hear as well in space?

We had difficulty hearing each other beyond twenty-five feet. Part of this was due to the noise level in *Skylab,* but some of it was probably caused by the thin air, as the atmosphere was about one-third as dense as air on the Earth's surface. There was no detectable change in the ear's ability to hear.

The shuttle's cabin pressure is about the same as on the Earth's surface, and the compartments are smaller. Any difficulty in hearing each other is probably due to the high noise level. [See Question 43.]

85. Did you have trouble talking? Did your voice change at all?

No, we didn't have any difficulty talking, even though the air was much thinner than air on Earth. We did not notice any change in voice pitch like the "Donald Duck" effect that deep-sea divers notice from breathing a thin or low-density gas such as helium.

86. Do things taste and smell the same?

There do appear to be some slight changes in the sense of taste, smell, appetite, and food preference. I didn't think the taste or flavors were as strong as on Earth. Part of this may have been caused by the nasal congestion I had.

On our flight, we repeated taste and odor tests that we had done on Earth before the flight. The results were different for each person, and no consistent patterns were determined.

There does seem to be an increase in the use of condiments. Astronauts who normally avoid hot and spicy foods have been observed to use Tabasco sauce quite liberally after a few days in space.

87. How did you tell time in space?

There is no "natural" time zone for space. However, it is important to use a single time standard so that mission planners and investigators from different countries can avoid confusion in scheduling activities. This standard is referred to as GMT [Greenwich mean time], the time at Greenwich, England. If you wish to include the date and year as well, then the time is referred to as UT [Universal time].

To tell time during the mission, we wore commercial wristwatches and also had several electronic clocks in the spacecraft. Our daily routine [work/sleep times] was based on central time in the United States. Our wake-up call came at 6:00 A.M. U.S. central standard time [CST]. Of course, day and night periods in orbit changed much faster than on Earth [sixteen sunrises and sunsets every twenty-four-hour Earth day]. I wore a wristwatch on each arm, with one set on GMT and one set on CST. All the *Skylab* clocks displayed GMT.

Experiments and work tasks were scheduled on GMT. However, the work shifts at Mission Control and our meals were scheduled on the CST cycle; it was helpful to have both times available.

For shorter missions like *Apollo* and the shuttle, mission elapsed time [MET] may be used. MET is useful when it is important to schedule events from lift-off [launch] time. If the lift-off time changes due to a delay, the MET schedule [flight plan timeline], based on the actual lift-off time, will still be usable. For example, an event scheduled for MET 2:16 would be two hours and sixteen minutes after lift-off.

Having Fun in Zero Gravity   Astronaut William R. Pogue (bottom) gives Scientist-Astronaut Jerry Carr a lift in weightlessness during the third Skylab Mission in 1974.

88. How can you tell up from down?

In weightlessness, there is no up or down insofar as your body feel is concerned. However, we did prefer moving to a position so that things "looked" right side up to the eyes. It was amusing to watch one of the other crewmen looking out the window toward Earth. He would always move his head or body around until his head was "up" facing the horizon.

It is also better to use a common "up/down" [architectural] positioning of workstations in a given area. A common orientation of control and display panels helps prevent mistakes in display interpretation or in operating switches and controls.

89. Does it make you dizzy when you do tumbling and acrobatics?

Yes, in a way. Doing rapid rotations or tumbling gives you a strong giddy, dizzy feeling like you get on a ride at an amusement park. The strange thing about it is that the dizziness isn't disorienting or the least bit nauseating after the first few days in space. It's a fairly powerful sensation with no ill effects. We still don't quite understand it.

90. What is a space suit made of?

The *Apollo* and *Skylab* suits were similar and had about fifteen layers of material. Starting from the inside, the suit contained the following materials:

a. A soft comfort layer of heat-resistant material called Nomex
b. A gastight bladder of cloth-reinforced rubberlike material, neoprene-coated nylon
c. Many layers of insulating materials

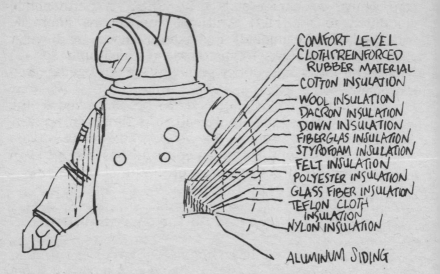

COMFORT LEVEL
CLOTH-REINFORCED RUBBER MATERIAL
COTTON INSULATION
WOOL INSULATION
DACRON INSULATION
DOWN INSULATION
FIBERGLAS INSULATION
STYROFOAM INSULATION
FELT INSULATION
POLYESTER INSULATION
GLASS FIBER INSULATION
TEFLON CLOTH INSULATION
NYLON INSULATION
ALUMINUM SIDING

A SPACESUIT MAY HAVE 15 LAYERS

d. A protective outer layer of glass fiber and Teflon cloth to protect against small meteoroids and fire

The bubble helmet was formed of a high-strength plastic called Lexan. Altogether it weighed about forty pounds.

These suits were customized [tailor-made], expensive, and had a limited service life.

For the shuttle a new suit was designed so it would be less expensive, enable "off-the-rack" fitting for a crew member, and would last a lot longer (fifteen-year service life with proper maintenance). It is referred to as the EMU [extravehicular mobility unit] and consists of three major components: the upper torso (also referred to as the HUT, or hard upper torso), the lower torso (or trousers), and the portable life-support system [PLSS, pronounced "pliss"]. The HUT is constructed of aluminum and the "trousers" are similar to earlier suits. The PLSS is permanently mounted to the HUT, with all connections internally attached. This eliminates the external hoses and water connections that were required on the earlier lunar surface suits. The lower and upper parts are joined together at a waist ring.

The shuttle suit does not have to be connected to the spacecraft with a hose [umbilical]. It is similar in appearance to the suit used by the *Apollo* astronauts on the moon but has many improvements and has enough oxygen and electrical power for seven hours of operation.

91. What do astronauts do for entertainment?

On *Skylab* we had an entertainment kit which included books, playing cards, squeeze-type hand exercisers, some balls, a Velcro-covered dart board with Velcro-tipped darts, three stereo tape players with headsets and cabinet speakers, and a pair of binoculars. We each selected our own music tapes and books in advance of our mission, and these were sent up in *Skylab* when it was put into orbit.

We used the tape players, binoculars, and books more than any of the other items. The darts didn't work too well. Their fins were small, and because of the thin atmosphere in *Skylab* [one-third the atmospheric pressure of Earth], they wobbled around when they were thrown. The second *Skylab* crew tried enlarging the fins, but it didn't help much.

Sometimes entertainment opportunities came up quite unexpectedly. We had dry-roasted peanuts in small cans with thin plastic covers. The covers had crisscross cuts to allow us to reach in with our fingers to pull out the peanuts. Occasionally, a peanut would float out of its container, and as we made our way through the space station, we would notice it drifting and tumbling through the air. When this happened, we would get against the wall, open our mouths wide, shove off toward the peanut, and try to capture it with our mouths like a fish. Sometimes we were lucky enough to catch it on the first attempt, but usually we would bump it, which would send it twirling off away from us.

I found that looking at the Earth with the binoculars was the most pleasant form of off-duty activity. The Earth was fascinating, and I never tired of looking at it. Next to this, I enjoyed the books and music the most. Floating acrobatics were also a lot of fun.

Astronauts on the space station will have video players that will support a wide variety of recreational video activi-

THEY BOUNCED ALL OVER THE PLACE

ties. In addition to movies, two-way family communications, etc., they should be able to support electronic games. Cosmonauts have had two-way video capability for several years.

92. Could you play basketball in space?

We didn't, but that's an interesting possibility. I don't think you could play basketball in the normal way. Dribbling would be a real challenge. We had three small balls in the recreation kit, and we played with them occasionally. When we threw them around, they bounced all over the place because of weightlessness. The hardest part was trying to find the ball when you were done. I do think it would be possible to think up an interesting ball-and-hoop game for weightlessness or, perhaps, three-dimensional billiards.

93. What was it like on the rocket going up?

When the engines fire up, they sound like muffled explosions, and there are a lot of noises from engine pumps and liquid fuel [propellants] surging through the large pipes [feed lines]. Early in the launch, there is a lot of shaking and vibration. As the rocket picks up speed, it lurches, twitches, and wiggles from thrust pulsations and abrupt swiveling [gimballing] of the engines to steer the rocket. It feels like being on top of a long wienie that's being shoved through the sky. There is a lot of swishing air noise as the rocket picks up speed. After the speed becomes supersonic, the swishing air noise suddenly stops, and then you can hear the noises from deep within the rocket, mostly creaks and groans.

When the fuel is burned out of the first stage, it is discarded or cut loose by explosives [pyrotechnics]. This is really an experience—it sounds like a train wreck. There are banging noises and flashes from the explosives and from the little rocket engines that pull the spent stage away from the tail of the next stage. You can also see a lot of metal pieces flying away and twirling lazily around the rocket. Staging only takes a few seconds, but it seems much longer. Then the engine of the next stage fires, and you're off toward orbit.

As long as we launch crews with rockets, I think it will be an exciting experience. Perhaps when we develop an aerospace plane, one that can take off from a runway and fly into orbit, it won't be quite as dramatic.

94. Were you ever really scared?

One of the NASA doctors asked me if I was scared at lift-off. I told him I didn't feel particularly excited. He said, "Well, that may be true, but your heart rate went from 48 to 120 at lift-off." I still don't think I was scared, but I must have been excited.

An old cliché often quoted by pilots is that "flying consists of hours and hours of boredom interrupted occasionally by moments of stark terror." I've also heard test pilots and astronauts say that they don't experience fear, merely varying levels of anxiety or apprehension. I think some of them are telling the truth, but I also think they are expressing a distinction between fear and panic. A person can be genuinely afraid but, through discipline, self-control, training, experience, and professional competence, can still function rationally and effectively to cope with problems or emergencies. It is also true that a person can become conditioned to react with some degree of detachment when faced with serious and life-threatening situations, particularly when they occur within the individual's area of professional expertise. If anyone can claim such self-control, I think it would be experimental test pilots, particularly those that are still around to talk about it. However, I believe anyone is capable of experiencing fear—and a high level of concern for personal safety, prestige, or professional status. The key is to avoid panic at all costs, and this is best achieved by being well trained. Also, I don't equate excitement with fear. A bit of controlled excitement really gets the mind alert and working.

All three of us got a real shock just after we had jettisoned our service module after the deorbit thrusting maneuver. The attitude required for this jettison task was way off the attitude required for reentry, which was only minutes away. At this point Jerry was to maneuver to the entry attitude using the attitude control rockets on the reentry module [command module]. When nothing happened, I looked over and saw him moving the controller, but it was obvious the jets weren't firing in response to the command. I yelled "Go direct!" [a backup method], but he had already switched over and the thrusters fired. That was the sweetest noise I'd heard in three months. We were already a bit concerned, because one of the two attitude rocket systems had already failed and had been turned off. During that brief period I would say that I was scared. If we hadn't been able to get to the right attitude with the heat shield facing forward for reentry, it would have gotten very warm inside.

95. What is the greatest fear in space?

The greatest concerns are fire and loss of air from the spacecraft. We had fire extinguishers on *Skylab,* and we also had emergency procedures to follow in the event of fire or rapid loss of pressure.

One day the fire alarm sounded as I was exercising on the bicycle. It was a blood-chilling sound, and I never liked to hear it, even when we were testing the alarm system. It turned out to be a false alarm, but it took me about half an hour to check out everything and determine that it was a false alarm. Our fire sensors detected ultraviolet light [emitted by fire]. However, sunlight contains ultraviolet light, and the false alarm was probably due to reflections of sunlight that came through a window in our wardroom.

When the alarm had sounded, my back was covered with sweat from the heavy exercise. I leaped off the bicycle so fast I slung off several large globs of sweat that splashed and bounced on a wall nearby. After checking out the alarm I had to mop up the mess with a towel.

96. How do you know when the spacecraft is perfect for launch? What is the risk of going into space?

There is no such thing as a perfect spacecraft or booster rocket. All equipment that is launched has flaws or imperfections. Some of these are known to exist, and other problems don't appear until the spacecraft is in space. The people who prepare the hardware for launch must consider all the known problems and decide which are safety-critical; that is, which problems may be a significant risk to crew safety or prevent accomplishing the main jobs scheduled for the mission. Problems that are considered critical must be corrected before launch. Other problems that are minor or merely nuisance imperfections may not be corrected.

Undetected problems or problems that arise during the mission are handled various ways. Things [functions] that are critical are duplicated within the design of the spacecraft so if one system or thing fails, there is another, independent system to provide that same function. For instance, the shuttle has five independent computers to perform the calculations required during launch and reentry. If one fails, there are four others left to do the job. Another way to correct for a failure during the mission is to repair the problem.

Other risks or hazards of space flight may be caused by human failures. People may make a bad decision or may make a mistake in performing their job. The *Challenger* disaster was caused by a bad decision. This cost the lives of seven astronauts. I made many mistakes during the three-month *Skylab* mission. Sometimes they were minor and merely required repeating the job. On other occasions the mistakes caused the loss of film or data.

The space program uses many ways to try to avoid human errors, such as using one group to check the work of another group. These checks and verifications are usually very effective in identifying problems. In the case of the *Challenger* explosion this system failed to operate effectively. Basically, NASA and industry people were under a lot of pressure to meet launch schedules, with insufficient resources to do the job. They failed to keep safety considerations as the number one priority and made a decision to go ahead with the launch even though good judgment would have dictated otherwise.

Many changes have been made in the NASA management system in order to assure that groups concerned with safety have a strong voice in making decisions when any question arises. However, there is absolutely no way to guarantee that space accidents will not happen again; it simply isn't possible. It is possible to enforce a system to reduce the probability of another space failure and this is what NASA has done. There will always be an element of risk in any space flight.

97. What kind of problems did you launch with?

Two days before our scheduled launch an inspector discovered cracks in all eight tail fins of our Saturn rocket booster. This was considered a serious problem, and the launch was delayed a week while the fins were replaced. After the fin replacement another inspection revealed cracks in the interstage trusses that connected the two stages of the booster rocket. These flaws were called stress corrosion cracks and could only be seen with a magnifying glass. A meeting was held to consider the risk, and it was decided that these latest cracks did not pose a hazard.

Our launch had already been delayed a week for the fin replacement, and when we were told about the stress corrosion cracks, I remarked to Jerry Carr that we ought to name our booster rocket "Humpty-Dumpty" because they were finding so many cracks in it. Later, Jerry casually mentioned our proposed name to the launchpad manager who had been working around the clock supervising the repair work crews who had accomplished the fin replacement and inspection work. He didn't appear to find the remark amusing.

The next morning we were atop the rocket in our spacecraft waiting for launch. In between the many checks we were confirming with the launch director, he would read "good luck" messages from the different teams that had participated in our training and launch preparation. It was rather nice and helped pass the time. Finally, the launch director said, "I have one final message," and Jerry Carr said, "Go ahead." The launch director read it slowly: "To the crew of *Skylab 4*, Good luck and Godspeed. Signed: All the King's horses and all the King's men." We had a good laugh and thanked the repair team for all their hard work. It somehow seemed reassuring to know they had a good sense of humor.

98. Did you get homesick?

I didn't get homesick in the strictest sense; that is, I didn't fret about it. We all missed familiar faces, but it wasn't a real psychological or emotional problem, because we were so busy. We were also mentally prepared to stay in space for eighty-four days; that was our goal, and we were psychologically oriented to the eighty-four-day mission.

There was also an interesting personal reaction I observed in myself as we neared the end of the mission. There was some consideration of extending our mission for two more weeks. We didn't hear too much about it on board, but we were unanimously opposed when some veiled suggestions were made. We quickly pointed out that we were out of food, which was technically correct. However, we probably could have scraped together enough spare meals from leftover food items to last two weeks. I remember thinking that extending the mission was a lousy idea. I had stayed up for the agreed-upon time, and that was that. In retrospect, I think my reaction would have been different if the approach had been different and had been made by the right person or if some emergency or operational problem had required it.

WE COULD HAVE SCRAPED TOGETHER ENOUGH SPARE MEALS

We all missed being around people, particularly family and friends. When we got back, it was very satisfying just to have a lot of different faces around.

There is an ongoing study of the problem of long-term isolation and confinement. The Soviets have much more space experience with this issue than we do and appear to have a lot of data based upon professional observation/evaluation of their cosmonauts during and after their missions. Analysis of the experience gleaned from Antarctic "winter-over" teams and from nuclear-powered submarine crews has provided many valuable insights into the problem of isolation and confinement. This becomes a critical mission-planning issue when considering a manned flight to Mars. A Mars mission will last from two and one-half to three years, most of which is spent in transit [outbound and return legs].

99. What was the hardest thing to get used to?

The head congestion or stuffiness. This was a minor problem on most space flights, but I seemed to have it a bit worse than my two fellow crew members. In space the sinuses don't drain as readily as they do on Earth; there's no postnasal drip in space.

100. Didn't you get bored on such a long mission of eighty-four days?

We were kept very busy, so boredom wasn't a problem. Many of the experiments we operated required a lot of operator input, and the creative involvement led to a high level of individual job satisfaction. I believe this was a key factor in preventing boredom. If your tasks are all strictly mechanical, only following inflexible procedures, then you very quickly lose interest or question your value in the effort.

I would have really enjoyed having more time to relax and look out the window. The U.S. experience with this problem doesn't even come close to the wealth of data the cosmonauts have provided. From informal reports, it appears that they do have a problem with personal motivation during their long missions. Soviet observers have commented on the progressive decrease in the time the cosmonauts can tolerate operating experiments during their workdays.

101. Was it possible to get any privacy?

Yes, we each had a separate sleep compartment with a fabric door and we would use this to read as well as sleep. Also, *Skylab* was quite large, so it was possible to get privacy by going to another part of the space station.

One rather amusing aspect of our sleep compartments was caused by the Velcro strips that latched our doors, which were really fabric sheets. When one of us got up at night to go to the bathroom, we opened the door by pushing open the Velcro strip. It sounded like someone was ripping open a shipping crate, and it frequently awakened the others. One night Jerry woke me, so I got up to look out the window for a few minutes. He was already there; we were over the Pacific Ocean somewhere and finally figured out we were flying over the Society Islands. We saw Tahiti, but it was mostly under clouds. After watching the coast of Chile come up, we gave it up and went back to bed.

Space station *Freedom* will provide individual crew quarters to provide privacy. A cosmonaut visiting the Johnson Space Center was shown a mock-up [model] of these crew quarters and expressed alarm when he saw that the quarters could be closed off with sliding doors. He expressed the view that crew members shouldn't be allowed to shut themselves off from the others. Perhaps it's a difference in the social or cultural habits, or it may be that we have something to learn when we begin staying up for longer periods.

102. How did your crew get along together?

Just fine. We were usually very busy, and there were so many problems with equipment that we had to help each other often. We had a good team spirit.

The only time the team spirit broke down was when I tried to draw a blood sample from Dr. Gibson. We had to do this every two weeks; Ed Gibson would draw Jerry's sample, Jerry would draw mine, and I was supposed to take one from Ed. Ed's veins were very small and I was probably the most inept at performing this task. After trying about ten times and causing Ed a lot of grief, he took the syringe away from me and drew his own sample. After that he never came back to me; he had Jerry do it, and it was never mentioned again.

103. Did you ever get mad at each other or have fights or arguments?

We didn't have any fights, and there was only one argument that I can recall. It had to do with a change in procedure, and the instructions were very vague. We resolved this by trying the procedure to see if it worked. We never got truly angry at each other, but we were frequently upset with or had disagreements with some people in Mission Control. We were all trying hard to get a job done, so there was probably fault on both sides at one time or another.

I think I upset Ed Gibson one day by putting his ice cream in the food warmer and leaving his steak in the freezer. I really felt bad about it. He couldn't eat the steak because it was still frozen hard, and the ice cream had turned to milk. He had to dig out some leftover food to make a meal. There wasn't too much conversation at dinner that night. He salvaged the ice cream by refreezing it. In liquid form it had turned into a big hollow ball. The next day, after it refroze, he stuffed it full of freeze-dried strawberries and had the first strawberry sundae in space.

104. Are you still friends?

Yes. Jerry Carr and I work together on several projects related to the space station and several advanced studies and projects. I see Ed about once a year and talk to him on the phone several times a year. Ed Gibson is also working on the space station with a Washington-based company.

105. What would you do if another guy went crazy?

This is not a silly question. Isolation and confinement can cause severe mental stress in some people, and it's difficult to predict to whom it will occur and also the extent of irrational behavior. Our crew had talked with an individual who had witnessed one such derangement in an Arctic situation, and he gave us a good idea of the warning signs. There are a lot of symptoms well before the time a person might cause harm to himself or others. The first sign is surliness and a general tendency to be uncooperative; the next is withdrawal from others. After a period of reclusiveness, the person gradually becomes openly antagonistic and aggressive toward others. If I were to have this problem, I would expect the other crew members to use whatever means available, be it medication or physical force, to control me while preparing to make an emergency/precautionary deorbit and return to Earth.

For a long mission like a journey to Mars, it will not be possible to perform a quick return to Earth. Handling a mental derangement problem will be one of the issues requiring careful advanced planning for such missions.

106. What would you do if someone died?

As missions become longer and more and more people go into space, the death of a crew member in space be-

WHAT IF ANOTHER GUY
WENT CRAZY?

comes inevitable. For the space station, the plan is to store the body in a respectful fashion and return the deceased on the next shuttle flight. It has not yet [1991] been determined exactly how the body will be stored while awaiting return.

One proposed procedure would be to place the deceased in a suit or body bag and place him outside in an area permanently shaded from the sun. The cold temperature would preserve the body.

The use of a suit does not seem realistic to me. It would be no better than a body bag and a needless use of a suit which might be needed for an operational emergency. Another proposal is to place him in a freezer in the galley area. This doesn't appear practical to me. It would take up most of the freezer space, and the immediate presence of the body would almost certainly have a depressing effect on the astronauts' morale.

Burial in space, similar to burial at sea, is considered unacceptable in Earth orbit. First, because there would be a need to do an autopsy, and second, because the body would remain in orbit and pose a safety hazard. However, this may be the only option on a long interplanetary mission. The disposition of deceased astronauts on Mars missions has not been considered.

107. Did anyone get sick and vomit?

Yes. I threw up the first day of the flight. This was an unpleasant surprise, because, according to tests we took while preparing for our mission, I was the least likely one of our crew to get sick. I used a burp bag similar to those available on airlines.

About one-half of the astronauts feel sick the first few days in space, but after about three days this is no longer a problem.

108. Did anyone get ill enough to need medical care? Were you prepared to take care of medical problems?

None of us had a serious medical condition. The worst problems we had were skin irritations, bloodshot eyes, headaches, head congestion, and cuts from working on equipment in hard-to-reach areas. We had been given limited medical training, including the ability to treat broken bones and sew up cuts. Also, we could talk to doctors on the Earth to get advice. We could have even shown the

THERE IS NO UP OR DOWN

doctors the problems by using television which we could send down to Earth.

We could cope with nonserious problems, but in the case of serious injury or illness, we would have given the person emergency treatment and returned to Earth. For example, if a person had an appendicitis attack, we would have given antibiotics to control infection and brought him back at the first good opportunity. We had a heart needle and a tracheotome to treat urgent emergencies like cardiac arrest and throat blockage.

We had a small pharmacy that included decongestants for stuffiness, sleeping pills, motion-sickness pills, antibiotics for internal infection, and aspirin for headaches.

For space station *Freedom,* the plan is to have a physician on the crew or someone with paramedic skills. For a Mars mission, it will be essential to have at least one physician with a backup [physician or paramedic]; someone with dental training will also be needed. Several of the current astronauts are qualified physicians and also competent engineers or pilots.

109. How did you know what medicine to use?

We used our own experience for minor things such as headaches and stuffy heads. We did have to report any medications taken. For more complicated illnesses, we would have referred to our medical treatment book or consulted with the doctors on Earth.

110. What would you do if someone got a toothache?

First, we would have treated it with medication. If that didn't work, we had all the equipment and training necessary to remove the tooth. We weren't trained to fill teeth.

111. If you cut yourself, would you bleed?

Yes, you would, but the blood wouldn't drop off. It would collect in a ball over the cut. If there was enough blood, it would just spread out on your skin. We had tissues and bandages to clean and dress any wounds.

112. Did your position of sleeping, up or down, affect the fluid shift?

No. There is no up or down in weightlessness, as far as the body is concerned. The fluid shift we experienced was caused by muscle tension in the legs, which caused certain body fluids to move toward the head and upper body. It had nothing to do with our position.

113. What happens when you sneeze? Would it propel you backward?

This is an intriguing question. I didn't have that experience—that is, sneezing—when I was floating freely and able to respond to the sneeze. The air rushes out your nose at about a hundred miles per hour during a sneeze, so it would act like a small jet or rocket and would propel you upward and rotate you backward. It would be somewhat like what happens when you let go of a balloon when the neck isn't plugged. But, because the body is so much heavier, the motion wouldn't be nearly as great as the balloon. The body motion resulting from a sneeze would, most likely, be less than the effect caused by the spacecraft cabin air circulating against your body.

However, I performed a simple calculation based on my body size, and the result was that a sneeze would cause me to do a complete backward somersault in about twenty minutes and move me about five feet upward during the somersault.

114. What could you see? What Earth features show up best?

Most of the time we saw oceans and clouds, but on almost every orbit we were able to get a good view of some land areas. The Earth features easiest to identify were coastlines, large lakes and rivers, major mountain ranges, and desert regions. Often it was like looking at a map, particularly when looking straight down at cloud-free land surfaces.

When we looked straight down toward Earth, we could see a distinctive feature as small as a football field. Color or shading contrast and unusual shapes were particularly helpful in improving our ability to detect and identify features. We were able to see icebergs about a hundred yards in diameter quite easily because of the contrast of white ice with the dark blue sea.

When we looked at pieces of hardware in space, we were able to see them with much greater clarity because of the absence of air. We noticed this first during launch, when our escape rocket and spacecraft launch cover were jettisoned about fifty miles above the Earth. As the rocket engines pulled the cover off the front and away from our spacecraft, we were able to see an unusual amount of detail in the structure of the cover. It seemed as though we could see every rivet and join line in it. When we got into orbit and turned around to look at our booster and, later, when we closed in during rendezvous with *Skylab*, we noticed the same thing—an unusual ability to see minute detail. In fact, objects looked so crisply and sharply defined that we got the impression we were looking at a finely drawn animation display. It was almost unreal.

The Soviet cosmonauts have noticed a significant increase in their ability to detect subtle Earth features and also to distinguish color differences. From their reports, they begin noticing this improved capability sometime after one hundred days in space. They refer to this as "space sight" or "second sight." They have conducted controlled tests, and the claim appears to be legitimate.

View From Space: Northeastern US and Atlantic Coastline Long Island can be clearly seen as the spit of land jutting out on the right. The Hudson River appears as a dark band extending towards the upper left of the picture. To the lower left are the Appalachian Mountains. (See question 114).

115. Could you see the Great Wall of China?

Yes, but we had to use binoculars. It wasn't visible to the unaided eye. The first time I thought I had seen it, I was in error; it was the Grand Canal near Peking. Later, I was able to identify the faint line of the wall, which zigzags in a peculiar pattern across hundreds of miles.

The Pyramids?

No. I was unable to see them even with the binoculars.

Lightning?

Yes. It was most spectacular in the equatorial regions where thunderstorms covered thousands of square miles. We could also see lightning in thunderstorms on the horizon over fifteen hundred miles in the distance.

The Grand Canyon?

Yes. It was very easy to see and identify. The colors of the walls of the canyon were quite obvious. After snowfall had covered the northern and southern rims, the colors were most vivid.

The Golden Gate Bridge?

No. We could see San Francisco Bay, but I was unable to see the bridge, even though I knew where it was.

The Aswan Dam of Egypt?

Yes. It's quite large and stands out clearly against the desert terrain. The Nile River is also very easy to see for the same reason.

The "airfields" of ancient astronauts, as popularized in books and television programs?

No. We examined the Plains of Nazca, at the foot of the Peruvian Andes near the Pacific coast of South America, but were unable to see the patterns in the plains. I took several pictures of this area and there are some very faint patterns—squares with circles inside. This is not similar to any of the patterns shown in aerial photographs of the area.

Lights at night?

Yes. In industrialized countries the lights are not only visible but are quite bright. Cities and major highways are very easy to see at night.

The aurora: northern and southern lights?

Yes. The aurora was the only Earth feature in which I could detect motion. The aurora patterns are variable and appear as dusky yellow tubes, spikes, sheets, and sprays.

View From Space: Egypt and the Sinai Peninsula   Lake Nasser is clearly visible on the lower right side of the picture, as is the Nile running across from it to the upper left. To its right can be seen the Sinai Peninsula and the northern tip of the Red Sea. (See Question 115).